Safety Risk Management:
Preventing Injuries, Illnesses, and Environmental Damage
Fred Fanning, CFM, PMP, LEED Green Associate

This publication is designed to provide accurate and authoritative information concerning the subject matter covered. It is sold with the understanding that the author is not engaged in rendering legal, accounting, or other professional services. If legal advice or other expert assistance is required, the services of a competent professional person should be sought

The reader should no rely on this publication to address specific questions that apply to a particular set of facts. The author makes no representation or warranty, expressed or implied, as to the completeness, correctness, or utility of the information in this publication. In addition, the author assumes no liability of any kind whatsoever resulting from the use of or reliance upon the contents of this book.

i

DEDICATION

To my maternal Grandmother,
June Madonna Armstrong.

TABLE OF CONTENTS

Chapter	Title	Page No.
	Introduction	1
1	Applications	3
2	Process	8
3	Tools	14
4	Using Risk Management for Wash Rack Operations	21
5	Using Risk Management for Rear Detachment Operations	24
6	Using Risk Assessment Codes to Rank Order Work	27
7	Adapting the US Army Risk Management Process for Emergency Operations	31
8	Adapting the US Army Next-Accident Assessment	39
9	Using a Risk Management Approach to Public Sector Worker's compensation	46
	Summary	57
	Glossary	58
	Bibliography	60
	About the Author	63

ACKNOWLEDGEMENTS

I want to acknowledge all the hard work done by US Army Tactical Safety
Professionals to further Safety Risk Management. These men and women
work alongside our soldiers in peace and war to integrate risk management
procedures into the Army Decision Making Process to reduce the risk faced
by soldiers

.

INTRODUCTION

If you could see into the future, know where accidents might occur and how serious they might be, would you take steps to prevent them? I am sure you would. I do not have a crystal ball, but in this short book, I will explain a technique that can be used to consider the future and determine what accidents might occur. This is not fool proof because managing risk is subjective and based on individual judgment.

Risk management is a process that begins in the planning phase of any activity, continues through the execution, and ends with activity close out. It is an iterative process where feedback is always informing the process of identifying additional risks and learning the status of control measures. However, that is not how risk management started. To find that out, we must look at the history.

A former colleague of mine, Ken Proper, wrote a three-part series of articles on the history of risk management. In his first article, Proper (2007) said that recognizing risk started with statesman Pericles in Athens around 431 BC. Proper then went on to explain risks involved in the development of games of chance in the Middle East and noted that the Arabic word for dice is **al zhar** and later became known as a **hazard**. In his second article, Proper (2007) wrote about the use of statistics to determine risk in the insurance industry. In the third and final installment of the series Proper (2008) noted how scientists identified that one chance in one million was the acceptable level of risk. He went on to explain how this simple decision has cost the United States Economy millions of dollars. Also in his third article, he went on to explain that the term "risk management" was first used by Russell Gallagher of Philco Corporation around 1956. Lastly, Proper (2008) highlighted that later agencies of the United States required the use of quantitative hazard analysis and risk assessment.

Georges Dionne is another individual that provides great insight into the history of risk management. Per Dionne (2013), "modern risk management started in the 1950s with the concept of financial risk management evolving in the 1970s." He also stated that "in parallel with textbooks on pure risk management circa 1964, engineers developed technological risk management models" (Dionne, 2013). He took us further along the timeline than Proper and noted that "financial Institutions, including banks and insurance companies, intensified their market and credit risk management activities during the 1980s. Operations risk and liquidity risk management

1

emerged in the 1990s" (Dionne, 2013).

Thomas Coleman (2011) talked about two different forms of risk. He stated, "idiosyncratic risk is the risk that is unique to a particular form, and systemic risk is widespread across the financial system" (Coleman, 2011). Both types of risk are significant and are covered in this book.

One of the issues that Coleman addressed was defining risk down to a single number. People seem to prefer describing risk with a single number. I have heard this same complaint, but I have not found a single number to be very descriptive of actual risk. Coleman felt much the same. He said, "it is entirely misleading to think there is a single number that is the 'risk'" (Coleman, 2011).

In this book, I will address how to determine the probability that an event might occur compared with the severity of an event. It is important for you to understand both concepts to get a valid identification of the hazard. The book is broken into two parts. The first part addresses the basics of risk management The second part has examples of how risk management can be adapted to specific situations.

I will also give you some insight into various types of risk management and then settle on safety risk management. After that, I have a few chapters to show how to adapt the safety risk management process to a variety of situations. In the bibliography, I include a list of sources that I used to write this book. I encourage you to read these books and articles if you want to know more about the subject of risk management.

I cannot cover everything in this book, but I have tried to give you the most relevant information. If you would like additional information about this or any other safety subject, I encourage you to go to my blog at http://fredefanningauthor.com/ where there are other books that can help.

CHAPTER 1 –APPLICATIONS

Introduction

Georges Dionne (2013) said the goal of risk management is to "create a reference framework that will allow companies to handle risk and uncertainty." Dionne means that risk management can be used in a variety of applications. In this book, I focus on risk management with the goal of preventing injuries, illnesses, and damage to the environment.

Financial Risk Management

Georges Dionne's work focused on methods for the sector where risk management is used widely to control losses and exploit opportunities. With respect to the financial sector, Dionne (2013) identified five main risks:

- Pure Risk
- Market Risk
- Default Risk
- Operational Risk
- Liquidity Risk.

A subset of the financial risk is transferred risk, which is used extensively in the insurance sector. This subset is likely the largest sector in the United States using risk management. The insurance sector uses risk management to determine costs of insurance and at the same time limit the amounts paid out due to losses. The insurance industry also uses risk management to prevent accidents and the associated injuries, illnesses, and damage to the environment. The Risk Management Society has a great deal of information that you can find at the URL https://www.rims.org/Pages/Default.aspx. I encourage you to visit that site.

Operational Risk Management

The Aberdeen Group (2013) has identified "the concept for Operational Risk Management (ORM) is about creating a framework that will help executives, employees on the plant floor, and maintenance personnel understand and manage the risks impacting their organization, establish

3

processes to address these risks effectively, and implement procedures for corrective and preventative actions." This framework is the basis for operational risk management.

"Risk management usually focuses on the problem of measuring risk and the decisions that flow from that problem combining the uncertainty of outcomes and the utility of outcomes to arrive at the decision to manage risk." (Coleman, 59, 2011)

The US Army uses a version of Operational Risk Management. The Army originally developed a tactical risk management program that was initially used in Vietnam. This program was reinforced in the 1980s when I began working for them. The program focused on eliminating and reducing the risk to tactical operations. The goal was to prevent loss and damage of what the Army calls combat power. This included weapons, personnel, and equipment. The idea was that if combat power could be saved from the loss, it could be used to fight the enemy it was intended for. Army Field Manual 100-14 (1999) explains it this way:

Risk management assists the commander or leader in-

- Conserving lives and resources and avoiding unnecessary risk.
- Making an informed decision to implement a course of action.
- Identifying feasible and practical control measures where specific standards do not exist.

Risk management does not-

- Inhibit the commander and leader's flexibility and initiative.
- Eliminate risk altogether, or support a zero defects mindset.
- Require a GO/NO-GO decision.
- Sanction or justify violating the law.
- Eliminate the necessity for standard drills, tactics, techniques, and procedures.

The Army expanded the use of tactical risk management and renamed it operational risk management. The Army now uses risk management techniques to prevent off-duty accidents as well as in civilian operations. This has expanded the risk management program into an automated process that is quick and easy.

After hundreds of hours of formal and non-resident training, I was able to use the Army process well. I also had thousands of hours of experience on

which to base my modifications to the process. The Army five-step risk management program is the basis for what I now call safety risk management. Over the years, I have modified the process for use in many different sectors other than military operations.

Project Risk Management

The Project Management Institute (PMI) defines project risk management as "the processes of conducting risk management planning, identification, analysis, response planning, and controlling risk on a project" (A Guide, 2013). This process is different from other risk management processes I have seen because of how project managers conduct risk management. They integrate this planning with other processes that result in a project management plan. After that plan, the risks are identified. This identification is made similarly to other methods with one exception. Project Risk Management considers positive risks or opportunities. Project managers then conduct a qualitative and quantitative risk analysis to determine severity and probability that is followed by "developing options to enhance opportunities and reduce threats" (A Guide, 2013). Opportunities are the point where this method differs from others. Financial and project risk management programs both seek to enhance opportunities.

PMI uses a matrix that focuses on cost, time, and scope. These are the basic measurements of any project and are used to measure the impact of risks. They also use a moving scale to oppose these objectives. For example, if changes are made to the schedule there will be corresponding changes to either the cost and/or the requirements. This is the method used to compare severity and probability. The system is efficient and works well for managing risks in a project. To learn more about project risk management, go to http://www.pmi.org.

Enterprise Risk Management

Enterprise Risk Management (ERM) is a collective process that uses risk management to increase the probability of achieving an organization's goals and objectives by using all methods and processes of risk management in what most would refer to as a portfolio. A lot of people do not understand that ERM is not a risk management process, but the use of risk management across an enterprise.

Risk Assessment Codes

These codes are not a full process of risk management. They come from the use of the first two steps of the risk management process. The result is a code that identifies the severity and probability of a hazard to result in injuries, illnesses, or damage to property or the environment.

Failure Modes and Affects Analysis

Failure Modes and Affects Analysis (FEMA) is a risk management process that focuses on determining the potential failure. Engineers use FEMA in reliability engineering. This method also uses probability and severity.

Safety Risk Management

Safety Risk Management comes from the US Army's Operational Risk Management and focuses on identifying the probability and severity of hazards and control measures that will eliminate or control the hazard to prevent an accident. Using this risk management process helps us understand ways to reduce the probability that an event would occur or lessen the severity if the event does occur. The event referred to here comes in two types. Near miss is an incident that could have resulted in injury, illness, or damage to the environment. The accident is an event that results in an injury, illness, or damage to the environment. Risk management methods (Army Field Manual100-14, 1999) include:

- Hasty Risk Management. A quick use of the risk management process during an activity assessment. The steps in a hasty risk management are the same as a typical evaluation. Except they are done quickly on the ground right before an event occurs or after an activity has been changed, leaving the risk management that was done obsolete. Limit the use of hasty risk management to situations where preparation time is extremely limited, and you cannot avoid a hasty assessment.
- Deliberate Risk Management is the application of the safety risk management process. Risk management is used to formulate control measures for activities. Identify the tasks and associated hazards, and determine the base risk levels. Then identify control measures and the residual risks. Managers then decide to accept the risk or identify more control measures to reduce the risk. Deliberate is the preferred method of risk management.

Job Hazard Analysis

Job Hazard Analysis is a method of identifying the hazards associated with the tasks of a job or process. It consists of the following steps:

- Define the Task
- Identify the Hazards
- Describe the Hazard Controls to be Used
- Define the Risk Level of Each Hazard

I added defining the risk level of each hazard. Without this step, it is hard to determine which hazard to correct first or how severe the risk is. I have also used the risk assessment of the Job Hazard Analysis to recommend hazard pay for employees.

Work Control Permits

Some organizations use a work control permit instead of safety procedures. The work control permit is a step-by-step description of the work to be performed that includes the hazards with appropriate control measures. The overarching purpose of the permit is to determine the hazards and control measures. The permit also requires a signature, which helps ensure the control measures will get done.

Ergonomic Assessments

The Ergonomic Assessment can also be used to determine risks. This assessment is used to identify hazards in the man-machine interface and for each hazard define a control measure to eliminate or control the hazard. Add the probability that the hazards will cause or contribute to an injury or illnesses. Also, add the severity of each hazard if it creates an injury or illness. From these, you can determine the risk and then determine the residual risk after the control measures are implemented. You can also use these assessments to determine hazardous duty pay.

Summary

There are a number of risk management applications. Each uses the basic premise, which is to manage risks to prevent something negative or help something positive to happen. Tailor the process to the business sector you will use it in. In this book, safety risk management will be the primary focus, but it is helpful for someone practicing risk management to understand other processes that may inform the way they use risk management.

CHAPTER 2 –PROCESS

Introduction

With any discussion about risk management, it is important to recognize that it is not an exact science. "We must continually remember that the future is uncertain, and all our measurements only give us an imperfect view of what might happen and will never eliminate the inherent uncertainty of the future" (Coleman, 2011). Consistently using risk management helps you determine what hazards might occur in the future and how serious they might be. Coleman (2011) also noted, "Risk management is as much an art of managing people, processes, and institutions as it is a science of measuring and quantifying risk." There are broad procedures that can be applied to many different circumstances. In contrast, there are procedures tailored for circumstances. The objective of managing risk is not to remove all risk, but to eliminate unnecessary risk.

Experience

At one time or another you were exposed to risk management, whether you knew it or not. The most common exposure that comes to mind is insurance. If you have car insurance, an insurance agent determined the potential for you to incur a payout and charged you accordingly for the car insurance. Your accident risk is the result of the risk posed by your age, health, driving habits, and past driving history. The more likely you are to be involved in an accident, the more the insurance will cost you. Playing the lottery is another example of risk management in daily life. If you play the lottery, you can read the chances of winning on the ticket. When the jackpot is high enough, most people decide to play. Unfortunately, the amount of numbers played versus the number of combinations makes everyone's chances poor.

Rules

I would like you to remember that no matter how you use risk management, there are four basic rules you must adhere to:

1. Integrate risk management, do not add it on.
2. Make risk decisions at the proper level.
3. Do not accept unnecessary risks.
4. Always obey laws, standards, or codes.

Let's take a quick look at each of these rules. To do risk management effectively, you must integrate it into the process from the very start. If you wait until the end, it will cost too much to change the process to eliminate or control the risks. Thus, you will not manage the risks. Someone with the proper authority must make the decision to transfer, accept, avoid, or control the risks. You do not want a machine operator deciding to accept the risk that could damage equipment and prevent an order from being filled. You would probably want a manager to decide that important. When a risk can be eliminated or controlled without affecting the scope, cost or schedule of work, then eliminate or control it. Lastly, never try to use risk management as an excuse for not obeying laws, standards, or codes. I am not aware of any requirement that allows you to do that.

Strategies

Staying with the theme of four, I also want you to remember that there are also four basic risk management strategies:

- Transferring
- Accepting
- Avoiding
- Controlling

Let's take a quick look at each of these strategies. First, transfer the risk to someone else. The transfer is usually done by purchasing insurance. Second, accept the risks by doing the work without controls. Sometimes avoiding risk is the best strategy. Avoiding means changing procedures so that the risk no longer exists. Lastly is to control the risk. Control is often done by identifying the risk, implementing control measures, and accepting the residual risk.

The Process

I would also like you to remember the five steps of the basic risk management process (Field Manual 100-14, 1999):

1. Identify hazards during the earliest planning phases of the activity.
2. Assess hazards by identifying the level of risk involved in each task.
3. Develop controls and make risk decisions by eliminating hazards and obtaining appropriate approval.
4. Implement controls by integrating them into the appropriate sections of plans and procedures.
5. Supervise and evaluate the effectiveness of the control measures.

The risk management process that I like best is in Figure 1. Step 1 - Identifying Hazards and Step 2 - Assessing Hazards describe the risk assessment process. Many people do a risk assessment and claim to have done risk management. In fact, to do risk management you must also do steps 3 through 5, where the management of risk comes into play.

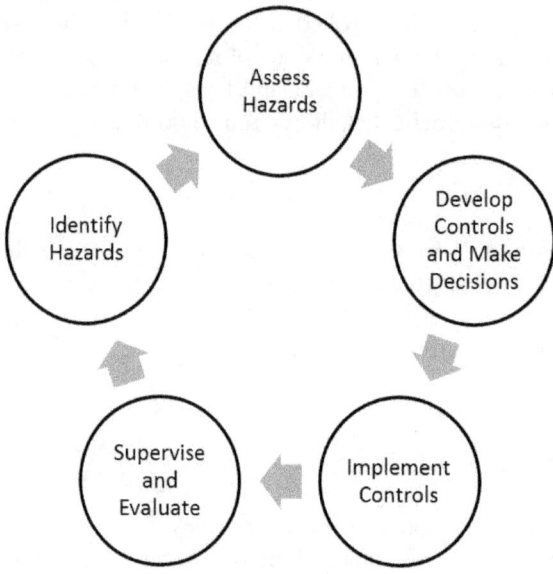

Figure 1 – US Army Risk Management Process (Field Manual 100-14, 1999)

These steps begin with identifying hazards; the rest of the steps are done in order the first time around. Then the process becomes a continuous loop that updates the process throughout the activity. For the rest of this chapter, I will explain each of the steps.

Identify Hazards

Identify hazards during the earliest planning phases of activity. All people participating in the event, whom I call stakeholders, should try to identify every possible hazard. It is best not to plan for a worst-case scenario. For example, when planning a trip, the likelihood that a commercial airplane would crash into the hotel should not be considered. The focus should be on an event with a more reasonable chance of occurring. In our example, a traffic accident on the way to the airport is more likely.

Use all available resources to identify hazards. First, stakeholders should identify hazards they know of in day-to-day tasks. To be effective, detection of hazards should go beyond the stakeholder's knowledge. What stakeholders do not know can affect the activity. Using detection resources, techniques, and tools provide a better chance of identifying hazards that might occur during the project. Consult with experts, engineers, planners, experienced employees, and supervisors. Reviewing standing operating procedures, standards, guidelines, previous incident reports, and Occupational Health and Safety Administration safety alerts for information about hazards can provide essential information. Also, reviewing related accident files and obtaining information from the organization's safety office can also be helpful. Use scenario thinking as another tool. Visualize the flow of the activity or task, the events that take place, and what could go wrong. Use other tools and techniques that are available. Use what works for you. I like to use brainstorming and scenario thinking to identify hazards.

Assess Hazards

The second step is to assess the hazards identified in step one. This step is more of an art than a science. No matter how you assess risks, it must be based on probability and severity. Probability is the likelihood that an event will occur due to the hazard. Severity is the expected consequence of an accident regarding injury, illness, and damage to the environment, or activity impairment. Activity impairment includes loss of property, equipment, and negative publicity. There are three steps to assessment. Step 1 is to determine the probability of each hazard identified. It is helpful to ask how likely it is that this hazard could cause a problem with the activity or task. Step 2 is to determine the severity of each hazard. It is helpful to ask how bad it could be. Step 3 uses a matrix to combine the results from severity and probability, see Figures 3 and 4 in Chapter 3. Determine a risk level for each hazard using the scale extremely high, high, medium, and low like the one in Figure 5 in Chapter 3.

Develop Controls and Make Decisions

The third step of the risk management process is to develop controls and make decisions. The fundamental objective of this step is to either eliminate the hazard or to reduce the risk to an acceptable level. In this step, you develop and select controls for each hazard, determine the residual risk of each hazard, determine an overall activity risk, and finally obtain a decision.

Controls must make sense. For example, controls that prevent a broken leg cannot cost several thousand dollars. Expensive controls are simply not cost effective. These controls must minimize the chance of an accident and at the same time maximize the chances of activity accomplishment. Continue the process of selecting controls until each hazard is eliminated or until benefits outweigh the risks. After selecting the controls, determine the residual risk for each hazard.

Residual risk is the amount of risk left after implementing control measures. To determine residual risk go back through the three stages for each hazard considering that the controls are in place. Once each residual risk has been identified, the next step is to determine the overall activity risk level. The overall activity risk is the single highest risk left after implementing controls, not an average of the residual risks. The decision to accept a risk is always a management issue. The person who makes the decision is the one who has authority from a determination by management for the specific risk levels.

Before you determine who accepts the risk in an organization, it is important to identify the risk culture. Risk culture includes determining how much risk the management will accept. Some people are risk averse, only accepting low levels of risk. On the other hand, you have people who will accept high levels of risk without a second thought. Most people are somewhere in the middle. No matter the risk acceptance level of management, it is necessary to know each person's level and adjust the risk management process to provide levels appropriately. A quick example is a factory manager who is risk averse. If we take him a high-risk project to maintain electrical switchgear while it is hot, he is likely to disapprove and be mad that his staff would even recommend something like this for his approval. In his case, the work should be done with the electrical power de-energized. That decision means it would never get to his level.

Implement Controls

Step 4 of the risk management process is to implement controls. The individuals involved in the activity or task implement the controls. When management approves the risk level they are also approving the control measures, that support that risk level. Supervisors of that organization ensure the details of the control are known by all employees involved. Each control must identify who will do what, by when. Each person participating in the activity must know what the controls are so that they do not skip or change one. Controls are accomplished through standard operating procedures or written and verbal directions given during activity or task safety briefings. Conducting rehearsals of the activity that include the use of the controls is a very effective method to ensure their integration. Rehearsals allow controls to become second nature or a usual way of conducting the activity. If you cannot mitigate risks to an acceptable level, the activity or task should not be done.

Supervise and Evaluate

The last and fifth step of the risk management process is Supervise and Evaluate. This action helps personnel perform to standard and demonstrate management's enforcement of the standards. Remember, for the risk management process to work, everyone must be accountable and responsible for his or her part. Supervision and responsibility go hand in hand. Every worker is responsible for performing to standard and performing controls. Every worker is responsible for recognizing unsafe acts or conditions and when possible making on-the-spot corrections. If workers cannot make corrections, they bring the hazard to the attention of their supervisor. The evaluation also provides essential feedback on how the process is working and if it is providing reduced risk for the organization. After-action reviews should be conducted with subordinates and superiors to assess the effectiveness of controls. At its best, evaluation should be performed during and after the operation. The key is that if the risk management process is not providing a service to the organization, you can adjust it until it does. Management should not expect to eliminate errors, flaws, or less-than-perfect performance. Risk management can be used to reduce the risk to an acceptable level.

Summary

The process does not stop with the five steps. As the planning and execution of the activity continue, it is essential for all personnel to continue to identify hazards they find and insert them into the process for assessment and control. Inserting continues until the activity is complete. The final step I recommend is to have an after-action review. This is a where all who participated in the risk management process review what was supposed to happen with what did happen.

CHAPTER 3 –TOOLS

There are tools that can help with the process of risk management. Most of these are not unique to risk management, and you may have used them in other situations.

Tools for Hazard Identification

The first tool is Brainstorming. This technique is used to gather ideas in a forum. A facilitator and recorder with a butcher paper or whiteboard join the stakeholders involved in the risk management effort. If the group is small, the facilitator and recorder can be the same person. Ask the stakeholders to identify hazards that exist with the activity or task that is being considered during the risk management process. The stakeholders should be told that each hazard will be written down. Allow only questions to clarify what the hazard is. Tell the stakeholders not to evaluate or comment on any hazards during the process. The facilitator should track time and stop anyone evaluating. Do this for about 45 minutes. The result is a list of hazards.

Another tool to help identify the cause of a hazard is to ask "why" five times. By doing this, you can go from the immediate cause of the hazard to its root cause. An example would be that the operator of the forklift who drove the forklift into a pole.

1. Why did he drive the forklift into the pole? Because he did not see the pole.
2. Why did he not see the pole? Because he could not see over the load on the forklift.
3. Why did he drive if he could not see over the load? He did not know he should drive backward if the load blocks his view driving forward.
4. Why did he not know to drive backward? Because he was not properly trained on the forklift.
5. Why was he driving if he was not properly trained? Because we needed a forklift operator in a hurry and just grabbed someone.

From that exchange, we know the root cause is that we did not accurately select and train the forklift operator, and thus he did not know to drive backward with a high load. Lack of knowledge led him to drive the forklift forward even though he could not see, causing him to drive into a yellow protective pole at the end of the shelves. It is always better to identify the root cause of a hazard than the most immediate cause.

Fault Tree Analysis is another tool that helps us identify the cause of a hazard. One example would be the brakes failing on a forklift. The brake failure is the start of the tree. We then identify branches of the tree that cause or contribute to the brake failure. The brakes may have failed because they were improperly adjusted, brake pads were worn, or the load exceeded the braking capacity of the forklift. The brakes may have been improperly adjusted because maintenance personnel did not perform preventive maintenance on the forklift or the operator failed to conduct pre-operational checks. The load exceeding the forklift capacity may have been caused by the operator not being properly trained, or the operator was trained, but not on this forklift. The tree provides a lot of information for hazards that might exist. See Figure 2 for an example of what a tree might look like.

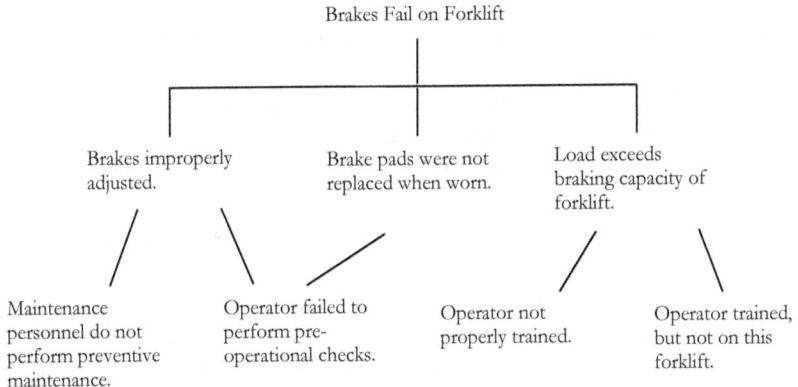

Figure 2 – Example Fault Tree Analysis

Tools for Hazard Assessment

Assessing hazards require an explanation to understand. Base this step on a comparison of probability and severity. Probability is the likelihood the hazards cause an event. Severity is the expected consequence of an event that results in an injury, illnesses, or damage to the environment.

Let's go back to Ken Proper's article for the Perspectives Newsletter. In this article, he stated, "risk management had its beginning in the science of probability which can be traced back to 1494 when Luca Pacioli proposed a problem concerning a game of chance (Proper, 2007)."

There are almost an infinite number of descriptions for probability, but I like to use five measures of probability to keep it simple. Figure 3 contains those five with a brief description of each.

P R O B A B I L I T Y	Frequent	Often occurs in the life of activity or may occur continuously.
	Likely	Occurs several times in the life of the activity, occurs at a high rate, but experienced intermittently.
	Occasional	Occurs sometime in the life of activity, occurs sporadically.
	Seldom	Occurs as an isolated incident during activity. Remotely possible, not expected to occur during an activity.
	Unlikely	Occurrence not impossible, but can assume will almost never occur in the life of activity.

Figure 3 - Probability Descriptions (adapted from Field Manual 100-14, 1999)

Thomas Coleman (2011) says, "Probability systematizes how we think about uncertainty and randomness." He also explains that because we apply the laws of probability to frequency we create belief-type probability or what we believe will happen.

We know events can have adverse outcomes. With probability, we estimate how many times an adverse event may occur in a total number of events. There is a second component to risk, which is the severity. We also need to know how severe an event might be if it is a bad event. We need to identify a range of severity from very minor almost de minimis through complete destruction or failure. The risk is the comparison of the probability that an event will occur and the severity of the impact. There are almost an infinite number of definitions for severity, but I like to use four measures of severity to keep it simple:

SEVERITY		
	Catastrophic	Death or permanent total disability of a person or damage to the environment.
	Critical	Permanent partial disability, temporary total disability more than three months to a person or significant damage to the environment.
	Marginal	Minor injury, lost workday accident, compensable injury or illness to a person or minor damage to the environment.
	Negligible	First aid or minor supportive medical treatment or no damage to the environment.

Figure 4 – Severity Definitions (adapted from Field Manual 100-14, 1999)

Once a description has been identified for severity and probability, a matrix is used to identify the level of risk. The matrix at Figure 5 is one that I have used a lot with great success. You make this matrix for the different descriptions of severity and probability. I have simplified this to Extremely High, High, Medium, and Low risk. These terms are relatively easy for most people to understand. The more complicated the matrix, the more difficult it is for everyone to understand and requires more training. You go to the point on the matrix where the word describing probability meets the cross section of the word describing severity. That word describes the risk.

Risk Matrix		PROBABILITY				
		Frequent	Likely	Occasional	Seldom	Unlikely
SEVERITY	Catastrophic	Extremely High	Extremely High	High	High	Medium
	Critical	Extremely High	High	High	Medium	Low
	Marginal	High	Medium	Medium	Low	Low
	Negligible	Medium	Low	Low	Low	Low

Figure 5 - Risk Assessment Matrix (adapted from Field Manual 100-14, 1999)

An excellent tool for determining control measures is a Hazard Control Matrix in Table 1. Identify the task workers will do to complete the activity, and then list the hazards, followed by a list of measures to control the hazard in column three. Try to provide a control measure for every hazard. An example might be a box truck driver hauling a load across the United States. The matrix for this trip might include the information in Table 1:

Hazard	Initial Risk	Control Measure	Residual Risk
Back up lights out on the truck	Moderate	Conduct a pre-operational check of vehicle lights and repair or replace any not working	Low
Tire might go flat	Moderate	Inspect tires and ensure spare tire and jack and tools are available	Low
Fatigue	High	Get eight hours of sleep before the trip starts; get sleep nightly on a two-day trip.	Low

Table 1 - Hazard Control Matrix

Another tool is Human Error Identification. Human Error is responsible for approximately 89% of all accidents. I like the system I was taught by the US Army. The human error involves a human somewhere at some time either by omission or commission causing or contributing to an incident or accident. Five inadequacies are responsible for the human error. It is important to consider these when assessing risk and considering control measures. The five are:

- Support failure
- Standards failure
- Training failure
- Leader failure
- Individual failure

Support failure is a lack of personnel, equipment or material, supplies, and services or facilities. Standards failure is that standards and procedures are not clear, not practical, or nonexistent. Training failure is that training was not correct, complete, sufficient, or to standard. Leader failure means that leadership is not ready, willing, or able to enforce standards. Individual failure is when an individual does not know the standard, lacks self-discipline, or is overconfident. Look at each hazard through the prism of these failures to assess how likely they are to contribute, and identify controls that would correct these inadequacies.

Tools for Risk Approval

The next step is to determine who accepts responsibility for the residual risk or what may be called controlled risk. I like to use a risk decision matrix. An example is in Figure 6, decision matrix based upon agreed-to positions within a company as to who approves of an identified level of residual risk. The low risk is accepted by an individual closest to the risk. As the risk increases to moderate, the approval level goes up, and when the risk gets too high it is usually the highest-ranking person for that level of the company that faces the risk. It is essential that someone accepts responsibility for the risk. When this is not the case, control measures are not always implemented. This puts the hazard back to the original level of risk with no control measures. Having someone accept risk by signing a document gets them to follow through with implementing control measures.

Risk Level Approval Matrix	Approval Level			
	Plant	District	Division	Enterprise
Low	Line Manager	Production Manager	Production Manager	Vice President
Medium	Production Manager	Production Manager	Vice President	President
High	Manager	Vice President	President	President

Figure 6 - Risk Decision Matrix (Fanning, 2002)

Tools for Implementing Control Measures

To manage risk, the control measures must be implemented. Otherwise, all you have done is to assess the risk. The person that implements the control measure should be identified by name. There also needs to be a due date assigned. Dates allow supervisors to follow-up and ensure that the work gets done. I like to use a resource matrix like the one in Table 2. I have completed a couple of rows as an example.

Hazard	Control Measure	Person Implementing	Date Completed	Date Verified
Dripping steam pipe connection.	Replace seal	David Wilson	3-15-2015	
Ventilator housing falling off	Fasten housing back to frame	Mike Johnson	3-17-2105	
Fan belt slipping on a massive fan	Adjust or replace belt as needed	Mike Johnson	2-18-2015	

Table 2 - Resource Matrix

Summary

This chapter focused on tools that can be used to identify the risk posed by hazards accurately. It is important to determine the probability that events occur and how much damage might result from the hazard. The fluid nature of risk management means that you can develop your categories and descriptions of probability and severity. To ensure the categories and descriptions are valid, they should be used several times with an after-action review to ensure they accurately measure requirements.

CHAPTER 4 – USING RISK MANAGEMENT FOR WASH RACK OPERATIONS

Introduction

Risk management is used extensively throughout the military deployment and sustainment phases of training and real life activities. I would like to expound on one of those uses here that controls risks when using wash racks to clean vehicles and equipment such as generators and tents.

Redeploying military units usually go through the process of washing and cleaning all vehicles and containers for shipment. There are a number of key players in the operation of a wash rack that form a cohesive team that focuses on improving the hazard control program and reducing risks.

Products

In this example, the primary hazard is freezing weather and risk management is used to control severe weather injuries, slips, trips, and falls. Risk management reduces the identified risks. This information is used to educate personnel who work on the wash rack daily, as well as transient personnel of deploying units, on the hazards and control measures to prevent injuries.

Two products were developed because of this risk management process that applies to wash rack operations conducted in cold weather. The first is the cold weather exposure guide at Figure 7. This hazard control matrix compares the temperature with cold weather equipment and clothing. By looking down the left side of the chart, you can locate the temperature after wind chill is figured in during which the wash rack operation takes place. The temperature at the wash rack helps identify the specific items of clothing required and the maximum amount of time a person should be exposed to the cold weather while operating a hose or steam cleaner. Anyone can use this quick and easy guide.

		Clothing Required	Exposure Time
F	50°	Normal clothing can be worn	No Limit
A	40°	Wet weather jacket and pants	4 hours
R E	30°	Wet weather jacket and pants Extreme cold weather clothing	2 hours
N H	20°	Wet weather jacket and pants Extreme cold weather clothing	1 hour
E I	10°	Wet weather jacket and pants Extreme cold weather clothing	½ hour
T	0	Wet weather jacket and pants Extreme cold weather clothing	¼ hour

Figure 7 – Cold Weather Exposure Guide (Evans and Fanning, 1999)

The second product is a wash rack authorization chart at Figure 8 that takes the risk of ice, snow, and temperature and identifies the approval authority that accepts the risk and authorizes the wash rack to operate. Identify the weather conditions in the right three columns that are present then move left to column one and determine the overall risk. Green is low, amber is moderate, red is high and black is extremely high. Column two identifies the approving authority.

Status	Risk {Approval}	Ice	Snow	Temperature
Black	Prohibitive {Brigadier General}	Sheet	Drift	<10°F
Red	Restricted {Colonel}	Patchy	Packed	10-32°F
Amber	Caution {Lieutenant Colonel}	Patchy	Light	32-40°F
Green	Normal {Major}	None	None	>40°F

Figure 8 - Wash Rack Decision Matrix (Evans and Fanning, 1999)

The decision to accept or reject a risk must be proportionate to the risk involved.

Summary

Cold weather injuries can occur when washing vehicles and equipment in extreme temperatures. Once a cold weather injury is sustained that body part will be affected every time it is exposed to cold weather again. This can limit the work this person can do. Some workers lose digits from fingers and toes due to severe cold weather injuries that have a long-term effect on them. It is important to take steps beforehand to prevent these injuries then try to cope with them after they occur. The procedures outlined in this chapter can be very helpful in preventing those injuries while still allowing workers to complete the work.

CHAPTER 5 – USING RISK MANAGEMENT FOR REAR DETACHMENT OPERATIONS

Introduction

A deployment is a process of moving personnel and equipment from a unit's home station to the training area or war zone. The military units always leave part of the unit behind. This element is called the rear detachment and includes the families of deployed service members. The risk management for the deployment should include the rear detachment. This example is unique to the military, but it provides another example of how risk management can adapt to the task (Fanning and Boggess, 2010).

Hazards Areas

There are six primary areas to identify these hazards in this activity, see Table 3.

Area	Definition
Geographic Location of Unit	Distance from the parent unit in driving time compared with unit type.
Geographic Location of Services	Distance from services in driving time compared with service type
Leadership Training	Kind of training provided and the level of Rear Detachment Command
Planning	Preparation guidance compared to the time for preparation.
Training/Exercise Schedule	Operational tempo guidance compared to type of training
Organized Family Support Group	Includes access and availability of privately owned vehicles, commercial transportation, telephone, family support group alert roster, and medical and dental facility.

Table 3 - Rear Detachment Risk Management (Fanning and Boggess, 2010)

Geographic considerations are more necessary for Army National Guard, Army Reserve, and organizations outside the continental United States. You measure the geographic location of the unit in driving time. You determine the hazard by comparing the type of unit with the driving time. The longer the driving time, the more risk involved.

Geographic location services include the distances from services measured in driving time. Services include a commissary, medical, dental, and post exchange. Location again measures the risk involved with operating a motor vehicle to obtain services. Leadership experience can cause or contribute to accidents because lower-ranking personnel normally have less knowledge, skills, and abilities than higher-ranking personnel. Units can offset this risk by providing training to personnel to prepare them for the duties and issues of rear detachment command.

You measure planning as in-depth, adequate, minimal, vague, implied, and specified. The general rule of thumb is to take 1/3 of the time allotted to conduct your planning and then give the subordinate unit 2/3 of the time for them to plan. Training and exercises use operations tempo compared to the guidance provided to determine the risk level.

The better organized and the lower the unit level with the support group, the less risk. Support groups also relieve lots of stress from family members knowing they are not alone at this tough time.

The level of risk identified is accepted or refused by the rear detachment commander. Determine the amount of approval authority in the early stages of the deployment planning process; see Table 4.

Risk Level	Unit Level			
	Squad	Platoon	Company	Battalion
Low	Squad Leader	Platoon Leader	Company Commander	Battalion Commander
Medium	Platoon Leader	Company Commander	Battalion Commander	Brigade Commander
High	Company Commander	Battalion Commander	Brigade Commander	Division Commander

Table 4 - Rear Detachment Risk Acceptance Matrix (Fanning and Boggess, 2010)

This risk acceptance approval authority should be agreed to in advance and documented. Base these levels on the knowledge, skills, and abilities of personnel at different ranks and describe the amount of risk each level agrees to accept.

Summary

Most times military units leave families behind. This can occur with the National Guard or even with emergency personnel. It is important to recognize that these families face more risk when an adult is deployed from home than if that person were to remain at home. These are unique circumstances that probably will not occur often. However, when they do adapt, this risk management method could come in very handy.

CHAPTER 6 – USING RISK ASSESSMENT CODES TO RANK ORDER WORK

Introduction

A risk assessment is the first two steps of the five-step risk management process. I do not recommend using a risk assessment code for anything other than a rank ordering work regarding hazard severity and probability. Hazards should be corrected on a worst-first basis using the risk assessment code (RAC) (Fanning, 2003). Stanley Kaplan and John Garrick (2001) say that risk assessment often involves three questions:

- What can happen?
- How likely is it to happen?
- What are the consequences if it does happen?

Codes

To determine the RAC, you identify the hazard severity category in Table 5.

Hazard Severity Descriptions
Category: I Description: Catastrophic Definition: Loss of ability to accomplish the duties or failure of duties of the organization. Death or permanent total disability of a person. Loss of equipment.
Category: II Description: Critical Definition: Severely degraded capability to conduct duties of the organization. Permanent partial disability, temporary total disability exceeding three months for a person. Major damage to equipment.
Category: III Description: Marginal Definition: Degraded capability of an organization to perform its duties. Lost day due to injury or illness is not exceeding three months for a person. Minor damage to equipment.
Category: IV Description: Negligible Definition: Little or no adverse impact on the organization to perform duties. First aid or minor medical treatment for a person. Slight damage to equipment.

Table 5 - Hazard Severity Categories (adapted from Field Manual 100-14, 1999)

Identify each level of severity by category, description, and definition of the result of being exposed to the risk. The best practice is to use the most common severity category that might occur and not a worst or best case scenario. After determining the hazard severity category, you must decide the probability category using the definitions in Table 6.

Probability Descriptions
Level: A Probability: Frequent, occurs very often, continuously experienced Single item: Occurs very often in the life of the equipment. Fleet or inventory of items: Occurs continuously. Individual employee: Occurs very often in a career. All employees exposed: Occurs continuously during an activity.
Level: B Probability: Likely, occurs several times Single item: Occurs several times in the life of the equipment. Fleet or inventory of items: Occurs at a high rate, but experienced intermittently. Individual employee: Occurs several times in a career. All employees exposed: Occurs at a high rate, but experienced intermittently.
Level: C Probability: Occasional, occurs sporadically Single item: Occurs some time in the service life of the equipment. Fleet or inventory of items: Occurs several times. Individual employee: Occurs some time in a career. All employees exposed: Occurs sporadically.
Level: D Probability: Seldom, remotely possible, could occur at some time. Single item: Occurs in life of equipment, but only remotely possible. Fleet or inventory of items: Occurs as an isolated incident in a particular activity or sometime in life, but usually does not occur. Individual employee: Occurs as an isolated incident during a career. All employees exposed: Rarely occurs within exposed population as an isolated incident.
Level: E Probability: Unlikely. Can assume will not occur, but not impossible. Single item: Occurrence not impossible, but can assume will almost never occur Fleet or inventory of items: Occurs very rarely (almost never or improbable) in an operation or the service life. Individual employee: Occurrence not impossible, but may assume will not occur in a career. All employees exposed: Occurs very rarely, but not impossible.

Table 6 - Accident Probability Category (adapted from Field Manual 100-14, 1999)

The best practice is to pick the most common severity category that might occur and not use a worst or best case description.

After determining the severity and probability categories, you can determine the risk assessment code by using Figure 9. Take the hazard severity category from Table 5 (I, II, III, or IV), put a finger on it in the left column of Figure 9. Run that finger across the row until it is in the column of the probability level (A, B, C, D, or E) the finger is now on the number comparing severity and probability. As an example, if the hazard severity category was a II and the probability category was B the number would be 2.

Hazard Severity	Accident Probability Risk Assessment Code				
	A	B	C	D	E
I	1	1	2	3	5
II	1	2	3	4	5
III	2	3	4	5	5
IV	3	4	5	5	5

Figure 9 - Risk Assessment Code Matrix (adapted from Field Manual 100-14, 1999)

The RAC would be II-B-2. This RAC means that an accident from this hazard could cause a critical situation that could significantly degrade the organization's ability to perform its mission. The result could be a permanent partial disability, temporary total disability of a person that exceed three months' time away from work. The result could also be major damage to equipment. This event is likely to occur several times in an activity or during the life of a piece of equipment. The result may occur at a high rate, but experienced intermittently with regular intervals, and often. An employee could experience the event several times in a career. Exposure to all the employees of an organization is at a high rate but experienced intermittently.

Summary

The definition of the RAC serves as the justification for why you do the work. Correct the worst first and prioritize. Using this method can significantly help explain which work orders need to be repaired first. In times when resources are hard to find, an explanation such as this can be worth its weight in gold. This method has been used with great success and I believes it will work for anyone.

CHAPTER 7 – ADAPTING THE US ARMY RISK MANAGEMENT PROCESS FOR EMERGENCY MANAGEMENT

Introduction

The US Army has identified two primary types of risk, tactical and accidental. An enemy creates tactical risk by using force whereas hazards create accidental risk. Field Manual 100-14 is the US Army document outlining this process and is on the World Wide Web.

The Army has lost more soldiers to accidents than enemy action in every war except Korea. This provides an incentive to prevent accidents through any means available. Through careful analysis, the Army determined that an accident rate can be reduced even in combat operations.

The US Army also makes trade-offs with risk management. For example, in Desert Storm US Army Commanders decided to take the battle to the enemy and push them out of Kuwait to reduce the tactical risk of military operations. To do this, the US Army moved rapidly extending the distances of the supply routes, which increased the risk of traffic accidents for vehicles traveling long distances. The US Army accepted more accidental risk for traffic accidents to reduce the tactical risk of the enemy causing injuries and damage.

The bottom line is that when US Army commanders control accidents, they give the Army back people, property, and resources to fight the enemy. The key to the US Army's success is the use of risk management to identify and control hazards before they cause an accident.

I believe that the US Army risk management process used to control military dangers can be adapted to do the same for emergency operations in other organizations. Tactical risks can include floods, fires, tornadoes, or even earthquakes. For example, a tornado poses a hazard to a community. If you do not respond to the tornado, you create a lost opportunity. This means you take no effort to protect or remove property or people from the path of the tornado. This inaction can result in property damage and people being killed and injured.

Risk management could work the same way in an emergency. For example, a city or emergency manager can increase the accidental risk for personnel and equipment sent to a flood to decrease the risk from the flood itself.

The US Army Risk Management Process

The US Army risk management Process is made up of five steps. The process starts with identifying the hazards and is iterative. Risk management starts the day a unit staff begins mission planning. The staff works risk management as part of the military decision-making process. They develop courses of action with the accident risk identified for each. The planning goal is to conduct the mission with the least risk possible (Field Manual 100-14, 1999).

The commander selects a course of action and approves the control measures that reduce the risk to the residual level. These decisions are fed down the chain of command to ensure everyone involved knows about the control measures that need to be implemented.

In the US Army, the Risk Management Process is owned by the commander. The commander's staff works the process with each member focusing on their specialty, which could be:

- Human Resources
- Operations
- Supply
- Logistics
- Engineering

As I noted earlier, the Risk Management Process is integrated into the military decision-making process. This integrates the management of risks into the process used to make decisions, which minimizes confusion and misunderstandings (Field Manual 100-14, 1999).

The US Army focuses on the hazard, not the risk. This is because the hazard hurts people, not the risk. Risks help commanders determine the danger they need to accept and approve. It is important for acceptable risks to outweigh the possible losses (Field Manual 100-14, 1999).

The US Army has tools that can be used to identify hazards. The METTC process is one of these tools. For METTC the military staff goes through each area that makes up the letters in acronym METTC (Field Manual 101-5, 2000).

- The letter M is for the mission.

- The letter E is for the enemy.
- The first T is for terrain and weather.
- The second T is for troops and equipment.
- The C is for civilians. In the Army, civilians are the people left on the battlefield who are non-combatants.

The next tool is the acronym METL:

- M is for Mission.
- E is for Essential.
- T is for Task.
- L is for List.

The military staff begins by looking at the mission or exercise they plan to conduct. First, they build in a description of the mission in the form of a list of tasks. Then they identify hazards for each task on the list. Through this method, they gain a considerable amount of information to review later.

There are also detection resources and techniques that can be used to identify hazards. These tools include brainstorming, experts, publications, accident information, scenario thinking, and rock drills.

There are also SMEs or subject matter experts. For example, the military staff can go to the safety professional for expert advice on risk management. There are other SMEs they can reach out to for needed information.

One of my favorite methods to identify hazards is to conduct a rock drill. The military staff places sand on a table. In the sand, they draw lines that represent buildings, roads, and highways. They put rocks in the sand that represent military equipment and personnel. Military staff moves the rocks around in the sand as the mission is executed. This educates everyone about the mission and the interactions of personnel and equipment. Through the process, the staff identifies hazards in the mission. These drills also save on gas and the potential for accidents of moving real people and equipment around on actual roads and highways.

Another aide is a risk management worksheet that can help military staff put all the information on one page that helps make sure that each step of the process is used. Each unit produces its sheet. The sheet must identify the operation name, date, and time of day. Identifying the tasks and putting them on the list is the meat of the process. The more tasks listed, the more comprehensive the analysis. Unfortunately, the more tasks listed, the more time it takes.

The US Army uses probability and severity to determine the risk.

Probability is the likelihood that an event will occur. It is divided into categories with the following definitions (Field Manual 100-14, 1999):

- Frequently – occurs very often, may occur continuously.
- Likely – occurs several times, occurs at a high rate while experienced intermittently.
- Occasionally – occurs sometimes, occurs sporadically.
- Seldom – occurs as an isolated incident, remotely possible.
- Unlikely – occurrence not impossible, can assume will almost never occur.

Severity is the expected consequence regarding the degree of injury or damage to the environment. It is divided into categories with the following definitions (Field Manual 100-14, 1999):

- Catastrophic – results in death or permanent total disability of a person, and major damage to the environment.
- Critical – results in permanent partial disability, or temporary disability of a person, and significant damage to the environment.
- Marginal – results in minor injuries or lost workday of a person, and minor damage to the environment.
- Negligible – results in first aid injury to a person and minor damage to the environment.

For every hazard on the list, the military staff identifies the definition of probability and severity that fits. This completes the first two steps of the US Army risk management process. At this point, the military staff has identified the risks associated with a mission and determined how risky they are, but they have not done anything to prevent or control the hazards.

The completed list gives them the initial risk without any control measures. Identifying controls and making decisions requires the military staff to look at the process through the lens of a cost-benefit tradeoff. If they spend time fixing the low hazards they can accomplish two things:

1. Probably get a lot of them fixed.
2. Not affect on the overall risk.

The military traditionally fixes the highest risks first and then moves down the list. Sooner or later they run out of resources to implement control measures. The military staff traditionally runs out of resources after fixing high hazards rather than running out of resources after fixing the low hazards.

The initial risk is the highest risk on the list before control measures. After they implement control measures, the residual risk is the highest risk remaining on the list. This is another reason to control or eliminate the highest risks first.

There are many ways to identify and develop controls. There may already be some controls in place before the risks are identified. For example, controls to prevent personnel from being exposed to a letter containing anthrax may have been established and published. Personnel is already trained on the procedures. These control measures are already in place and ready to be used.

The military staff is realistic about the risk management process. The military staff can make training or the mission so safe they cannot meet any of the planned objectives. Risk management should not be used to stop training or missions, but rather to enable training and missions to operate at the edge of danger.

On the risk management worksheet that the military staff listed all the tasks, they now list the control measures. Once they complete listing the control measures, they go back and determine the risk level with the control measures implemented. After this, they determine the residual risk, which is the highest-level remaining risk after control measures are implemented. They now must decide who makes the risk decision.

Typically the military commander makes the risk decision. Once the commander's staff identifies the residual risk, they frequently find that it is lower than the initial risk.

It does not matter if they reduce the probability or severity. Reducing either should reduce the overall risk level.

The final action is for the commander and unit leaders to ensure all the control measures are implemented so that they can be sure the residual risk will be achieved. All military personnel continues to identify hazards as they conduct the mission. These are raised to leaders who refer them to the military staff for inclusion into the risk management process.

Adapting the US Army Risk Management Process

The US Army risk management process can be adapted to reduce risks for people going to, from, and in and around emergency operations as well as risks posed by earthquakes, floods, or fires.

The standards outlined in US Army Field Manual 100-14 spell out the process for individuals to use in assessing the risk, determining which risks are controllable and how. It also explains that everyone is responsible for risk management. Adapting this standard is the best approach. There are a few basic concepts that will make the adaptation successful.

There should be a single person responsible for disaster preparedness and operations that make overall decisions. This person can delegate responsibility to the Fire Chief for fires, Chief of EMS for medical activities, and supervisors for placing personnel in response to the emergency.

They may also delegate responsibility for these same people to oversee risk management in their area. They should also delegate to people who work for them the ability to do the follow-through to implement control measures. The workers must be told what control measures are to be used and how to implement them. For example, a worker that is placing sand bags on the levy wall may also be given the control measure of warning indicator duties. When the water levy gets too high, he or she spreads the word back, so others know. Furthermore, if she or he hears a siren, they tell everyone to evacuate the levy immediately. The system works best when everyone has buy-in from the person accepting and approving the risks down to the person performing the actual emergency response duties.

Teaching risk management to all levels of the emergency management team is essential. The goal is to make sure that everyone involved in emergency operations is trained in risk management and prepared to make risk decisions to the very best of their abilities.

Leaders must never risk a first responder's life unnecessarily. I do not think they would do this intentionally, but without the full information, they could do it unintentionally. If emergency response leaders send people into burning buildings, floods or earthquake ravaged areas without being acutely aware of what could happen to those people they have risked their lives unnecessarily.

You should use the definitions from the Army Field Manual to ensure everyone speaks the same language.

The amount of experience a person has is directly related to their ability to manage risks. After being trained, people need to participate in the process often to get better at it.

Avoidance is the best method to control a hazard, but in emergency response, avoidance is almost never an option. You must always expose people, property, and resources to risk when responding to an emergency. People in emergency management know this.

The emergency management team should look at risk by comparing probability and severity the same way the US Army does. Working the two together reduces the risk to a level known as residual risk. What the emergency manager needs to know is:

- What are my risks?
- How can I control them?
- How much risk is left after I control them?
- Can I accept that residual risk?

The answers to these questions can be obtained by using the five-step process utilized by the US Army. The staff of the individual in charge of the emergency training, planning, or response should perform the risk management process. They should start risk management the day that an emergency response unit begins planning for an emergency or training. The emergency response staff should integrate risk management into the decision-making process used by emergency managers. The planning goal should be to conduct the emergency training, planning, or response with the least risk possible.

The emergency response staff should begin by looking at the emergency mission or training they plan to conduct. First, they develop a list of tasks. They then identify the hazards with each task on the list.

The staff can use the detecting resources and techniques that the US Army uses to determine the hazards including brainstorming, experts, publications, accident information, scenario thinking, and rock drills.

The emergency response staff should go to a safety professional for expert advice on risk management.

I also recommend developing your risk management worksheet for emergency response staff to use in the process of putting all the information on one page and help make sure the steps of the process are used. Your risk management work sheet should identify the operations name, date, and time of day.

Many organizations identify the tasks, hazards associated with them, and the risk involved. It is important not to stop with the first two steps of the risk management process. By identifying and assessing risks you have not controlled them.

You should continue by identifying controls for the hazards and determining the residual risk after controls are implemented. There are many ways to develop controls. Remember that there may already be some controls in place even before the risks are identified.

Your time is best spent controlling the highest risks first and then move down the list. Do not make a mission or training so safe that the emergency response staff is not able to meet the objectives set for the mission or training. Do not use risk management to stop missions or training, but rather to enable missions and training at the edge of danger.

The city manager or emergency manager must make the risk decision once their staff identifies the residual risk. The final action is for the city manager, or emergency manager to take steps to ensure all the control measures are implemented. The process should be iterative. All emergency response personnel should continue to identify hazards as they go about the mission or training. These are raised to their leaders who should refer them to the emergency response staff for inclusion into the risk management process.

Summary

I think you can see how the US Army risk management process can be adapted with very little change to managing risks for emergency operations. You do not have to use it exactly the way the US Army does, but use it in a way that works best for your organization's emergency planning.

CHAPTER 8 – ADAPTING THE US ARMY NEXT-ACCIDENT ASSESSMENT

Introduction

What would it be worth to know about the next accident that might occur in your organization? Knowing this information could save valuable resources and maybe even lies. There is something that can help workers determine their risk for having an accident. This is another tool developed by the US Army that is called the Next-Accident Assessment. The US Army has developed assessments for aviation personnel and leaders, ground personnel and leaders, and civilian employees. Most of these assessments are for military personnel conducting military operations. However, there is one that can be used in the private sector that is called the Next-Accident Assessment for Civilian Employees.

The tool is used by the worker, and he or she is not required to share this information with their supervisor. This tool provides better information if the results are shared. The power of this tool is demonstrated when action is taken to reduce the risks. In this assessment, the US Army refers to "chain-of-command." The private sector equivalent to that is the "chain-of-supervision."

The US Army based this assessment on the theory of Human Error Accident Reduction. This theory says that 80 percent of all accidents occur because of human error. The next paragraphs will explain the theory of Human Error Accident Reduction and how to use this assessment.

The US Army learned through accident analysis that 80% of accidents are the result of human error (Field Manual 100-14, 1999). This means that somewhere, at some time a human being though omission or commission caused or contributed to an accident. I believe that the same percentage is true for private sector operations. Furthermore, the person causing or contributing to the accident may not be anywhere near the accident when it happens.

For example, safety engineers evaluate equipment designs and purchases made by the US Army. Unfortunately, hazards are often built into the equipment. The job of the safety engineer is to minimize these hazards and still ensure that the equipment fulfills its purpose. Unfortunately, hazards remain after the equipment has been manufactured. The people who design the equipment will probably be nowhere near an accident when it occurs, but it could be a hazard they overlooked or failed to abate that causes the accident.

Systemic Failures

The US Army looks at five systemic problems around human error (Field Manual 100-14, 1999):

- Support Failure
- Standards Failure
- Training Failure
- Leadership Failure
- Individual Failure

A Support Failure is a lack of something. It should not surprise anyone that no one ever has all the resources they need to do their job. This absence of something may cause workers to take shortcuts that increase the risk of accidents.

You can manage standards away. In contrast, some standards may not exist; they may not be clear, or do not apply to your activity. This describes a Standards Failure. In this case, the standards may be created or modified by the training or a mission that is conducted. Soldiers often create a solution to resolve a problem as a part of the training or a mission that later becomes common practice or a written standard.

Training is often done poorly or not at all. The more thoroughly training is conducted, the lower the risk level that can be achieved. Training failures result from training that is not provided or performed poorly.
Leaders can create human errors. Leader failures occur when leaders do not step up and tell soldiers what they need to know to do something correctly. Leader failures were involved in every accident I investigated. The failure was because the leaders were not actively involved.

Lastly, there are individual failures. This is true even though individuals may not be the direct cause of the accident. They may do something immediately before or months before the accident that contributes to its occurrence.

Identifying human errors and controlling them keeps their impact to a minimum. The goal is to prevent human errors during training and emergency operations.

Adapted Assessment

To do the assessment, the worker answers several questions about themselves. Each answer comes with a number of points. The points are added up to determine the risk level as being low, medium, high or extremely high. The tool ends there, however – as with all risk management tools the next step of identifying control measures is essential. I have included my adaption of the Next-Accident Assessment for Civilian Employees in Field Manual 100-14 in this chapter.

The Next-Accident Assessment

The assessment is designed to provide individuals with immediate feedback on the risk factors they possess. The results are for their use only. However, it's an excellent tool that can be used by employees to take responsibility for their safety. Sharing the information with their supervisor increases the results of this assessment.

Will you cause the next accident?

This assessment will help you figure out, on our own, your chances of being the next accident statistic. To rate yourself, answer each question honestly and total the points to learn where you have risks. The theory of Human Error Accident Reduction says that human error is responsible for 80 percent of all accidents. These accident-causing mistakes happen for five reasons. Sometimes the individual who makes the mistake is at fault, and sometimes it is the person's organization or senior manager that is at fault.

The Next-Accident Assessment is based on the top five reasons for human error accidents involving self-discipline, leadership, training, standards, and support.

The assessment is for your awareness only. You do not have to share the results with anyone, but the whole organization can benefit if you do. Once you have completed the risk assessment, you can then act to correct or control the risk factors you identified. You can also identify actions you need for your organization to take to reduce your accident risk; this is the only information you need to share with your organization and supervisor.

Instructions

Answer the questions about yourself and assign points as directed.

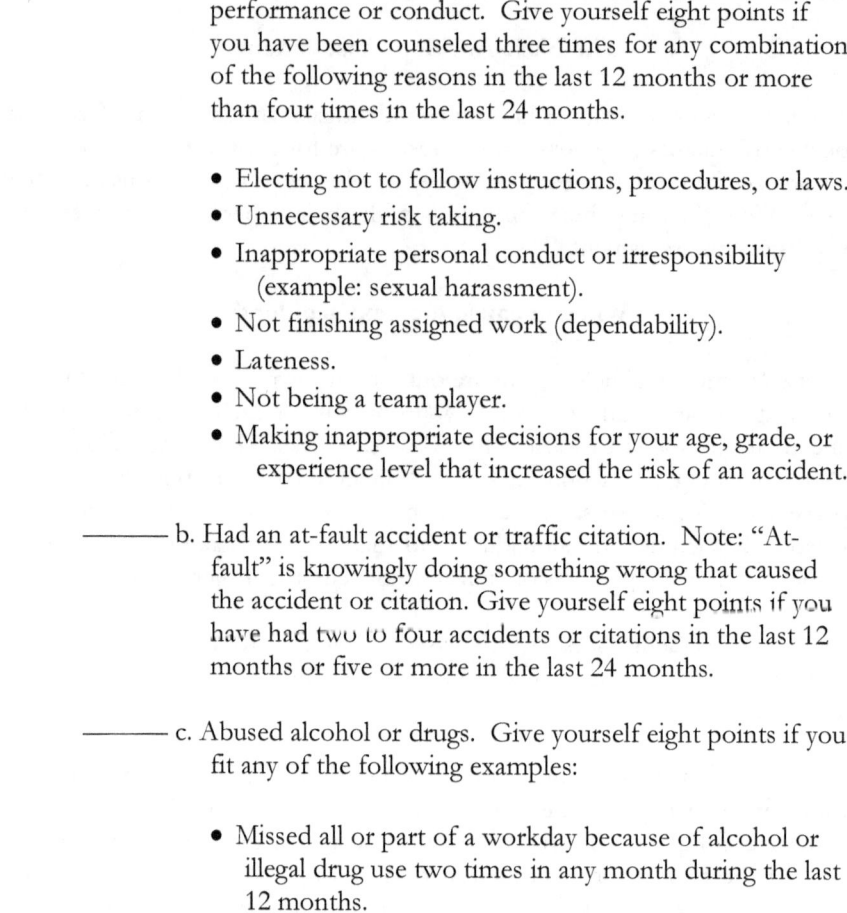

Points ——— Question 1. Self-discipline. You know the standard for performing your job tasks. You have been trained to perform those tasks to standard, but you frequently choose not to because of your attitude. This is a lack of self-discipline. The following are indicators of an undisciplined individual. Give yourself points for indiscipline if you—

——— a. Have been formally or informally counseled for poor performance or conduct. Give yourself eight points if you have been counseled three times for any combination of the following reasons in the last 12 months or more than four times in the last 24 months.

- Electing not to follow instructions, procedures, or laws.
- Unnecessary risk taking.
- Inappropriate personal conduct or irresponsibility (example: sexual harassment).
- Not finishing assigned work (dependability).
- Lateness.
- Not being a team player.
- Making inappropriate decisions for your age, grade, or experience level that increased the risk of an accident.

——— b. Had an at-fault accident or traffic citation. Note: "At-fault" is knowingly doing something wrong that caused the accident or citation. Give yourself eight points if you have had two to four accidents or citations in the last 12 months or five or more in the last 24 months.

——— c. Abused alcohol or drugs. Give yourself eight points if you fit any of the following examples:

- Missed all or part of a workday because of alcohol or illegal drug use two times in any month during the last 12 months.

- Been on duty while under the influence of alcohol or illegal drugs any day during the past months.
- Referred to Community Mental Health or other agency for alcohol/drug abuse evaluation during the past 24 months.

———— d. Received punishment. Give yourself eight points if you received punishment for any of the following in the last 24 months:

- Missed work.
- Crimes against property.
- Crimes of violence.

———— e. High School Diploma. Give yourself eight points if you do not have a high school diploma.

———— f. Sex and age. Give yourself eight points if you are a male under the age of 25.

Points ———— Question 2. Leadership. Your immediate supervisor is not ready, willing, or able to supervise subordinates' work and enforce performance to standard.

Give yourself 18 points if your supervisor fits either of the following examples:

- Your supervisor does not have sufficient technical knowledge, experience, or management ability to supervise adequately.
- Your supervisor tolerates below-standard performance, rarely makes on-the-spot corrections, does not emphasize by-the-book operations, or is reluctant to take disciplinary action.

Points ———— Question 3. Training. You have not received the training you need to perform your current job tasks to standard. This means that training was insufficient, incorrect, or not provided by schools, organization, or on-the-job-training experience.

Give yourself 18 points if either of the following examples applies to you.

- No proficiency training in the last five years.
- Not proficient in tasks outside your job series but required in the current job.

Points ———— Question 4. Standards. In your current job, you frequently perform tasks for which standards or procedures do not exist, are not clear, or are not practical.

Give yourself eight points if either of the following applies to you.

- Tasks in your job series either have no standards or procedures or have standards or procedures that are not clear or are not practical.
- Tasks outside your job series (other duties) assigned to you either have no standards or procedures or have standards or procedures that are not clear or are not practical.

Points ———— Question 5. Support. You frequently do not receive the support you need to perform your job tasks to standard. Shortcomings include type, capability, and amount or condition of support needed. Give yourself eight points if inadequate support was responsible for below-standard task performance two times in any month during the past 12 months. Examples:

- Personnel (not full crew, wrong job series, not trained to standard, etc,).
- Equipment (tools, transportation, safety, etc.).
- Supplies (fuel, water, parts, clothing, publications, etc.).
- Services/facilities (maintenance, medical, personal services, storage, etc.).

Scoring

Add up your points for all questions. Find where your score fits on the scale below to determine your risk of causing the next accident.

———— Total Points

Points	0-20	21-30	31-40	41+
Risk	Low	Medium	High	Extremely High

How to use the results?

You now know your risk of making a mistake that will cause the next accident and what the reasons will be. You can reduce your risk by acting to correct or control those reasons/faults that apply to you. You can control or fix some of them yourself; for others, you may need chain-of-command help. In the space below, identify at least one action you will take to reduce your accident risk. Also, identify at least one action you need the chain-of-command to take to reduce your accident risk (this is the only information you need to share with the chain of command).

Action (s) I will take to reduce my accident risk.

Supervisory chain of action (s) needed to reduce my accident risk:

Summary

Using this method can help workers identify where they have a risk in their work procedures. It is important to teach them this information. However, this information is not worth much unless action is taken to reduce the risk. That means that supervisors must work with them to identify actions and make sure steps are taken. Lives can depend on it.

CHAPTER 9 – USING A RISK MANAGEMENT APPROACH TO PUBLIC SECTOR WORKER'S COMPENSATION

This method was developed for an action learning event for an Executive Leadership Development Program. None of the participants in this learning event were safety professionals. The author, as a safety professional, developed the concept for and served as the executive sponsor for the team. The product of the action learning event was a report and a presentation made to the Safety and Occupational Council of an Executive Agency of the Federal Government.

There are a number of issues within the realm of workers' compensation that includes:

- Third party involvement
- Deceased compensation recipients who continue to receive benefits
- Surviving spouses of deceased compensation recipients who have remarried and are no longer eligible for benefits, but still receiving them
- Missing medical documentation needed to determine if an employee can return to work,
- Return to work of able compensation recipients
- Investigation of fraud

Background:

Federal civilian employees, as well as some contractors and volunteers for the federal government, receive workers' compensation payments in accordance with the Federal Employee's Compensation Act (FECA), Title 5 Part III, Subpart G, Chapter 81, Subchapter I. The requirements for this program are further codified in 20 Code of Federal Regulations, Part 1-199. The Federal Employees' Compensation Act provides workers' compensation coverage to three million Federal and Postal workers including wage replacement, medical and vocational rehabilitation benefits for work-related injury and occupational disease (FECA, 2008).

Compensation recipient's medical expenses are paid in full while income compensation is 66.67% of gross wages for employees with no dependents, and 75% for those with dependents (Injury Compensation for Federal Employees, DOL, 2007, 40).

Each federal agency incurs the costs of its own workers' compensation recipients but relies on the Department of Labor (DOL) to administer the FECA program. Agencies provide DOL with detailed information about compensation recipients. DOL then processes the claims and bills the agencies annually for reimbursement using "charge back" reports.

The "silver bullet" for worker's compensation is to return the compensation recipient to work. Unfortunately, there are a lot of people involved in the process as well as legal restrictions. There are as many ways to return a compensation recipient to work as there are claims. Some work and others don't; however, most are very subjective. The key areas that can be used to determine if an employee should be targeted for return-to-work are:

- Employee interest in returning;
- Amount of leave already taken;
- Employee's physical condition; and
- Reassignment factors.

What is needed is a methodology for ranking desirability of a return to work based on measurable criteria. This paper will elaborate on the information gathered in a literature search which led to 25 measurable criteria that can be used to measure the potential for success in a return-to-work effort. This chapter will then break these criteria into the four key areas listed above.

What is needed is a methodology for ranking desirability of a return to work. To support that methodology there must be measurable criteria identified that can be used to measure the potential for success in a return-to-work effort.

Discussion:

The author sponsored the team of individuals to develop a methodology for ranking desirability of a return to work based on measurable criteria. To support that methodology, the team also identified measurable criteria to measure the potential for success in a return-to-work effort. This method and supporting criteria would serve as a strategy for enhancing the return to work of ready and able employees who have been receiving worker compensation for an extended period. Figure 10 provides the model for the process of returning compensation recipients to work.

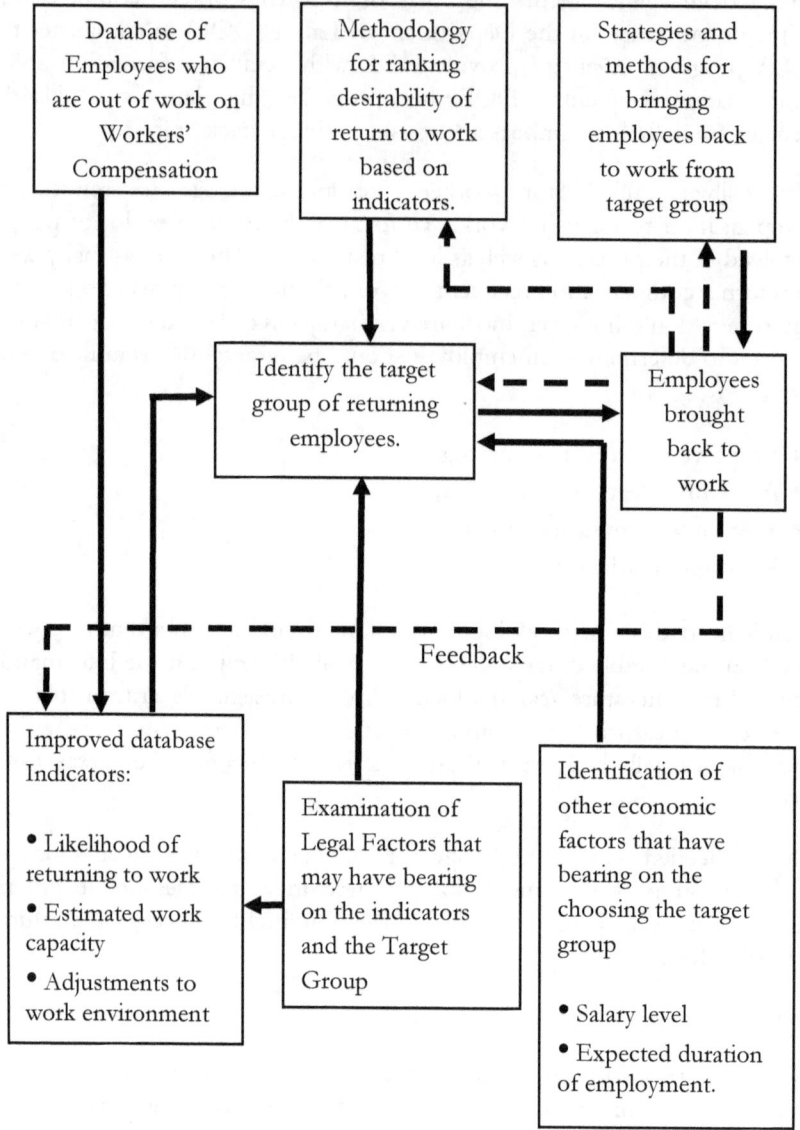

Figure 10 – Workers' Compensation Return to Work Process Map

Most Environment, Health, and Safety (EHS) offices have a database of claims that allows them to track the expenses; however, this database is normally not able to conduct the analysis needed to return compensation recipients to work (Ankel, et al., 2006). For that, the EHS office must first acquire or develop a single user-friendly database on the pool of compensation recipients who are on the permanent leave for worker's compensations. This database should include the following parameters for each compensation recipient listed in the original database.

- Likelihood of return to work, given what is known about the compensation recipient;
- The estimated work capacity (percent of full-time labor) that would be expected if the compensation recipient returns to work; and
- Categories of necessary adjustments to the work environment, e.g. work at home, special equipment, restriction on hours.

The analysis should result in a ranking of the desirability of workers' compensation recipients based on these parameters. It would also depend on other economic factors such as the salary level and the expected duration of employment. This will be called the target group because they will be targeted for a return to work. Once the target group has been identified the EHS office should work with human resources, supervisors, and budget officers to develop strategies and methods to bring these compensation recipients back to work, as displayed on the right side of figure 1. The consequences of these actions will then serve as important feedback into the various parts of the model, such as the development of improved database, the methodology for ranking desirability of return to work, and identification of the target group (Ankel, et al., 2006).

In addition to the new method of analysis, the return to work processes can be improved through more and better communication that leads to understanding among the physicians responsible for treating compensated injuries or illnesses, the offices within the employer organization that process the workers' compensation paperwork, and the actual workplace where the injury occurred and where the compensation recipient is likely to return. Supervisors must know the compensation recipient's ability to perform the "essential functions" of his or her job. To do this, the supervisor must know the physical and mental condition of the compensation recipient following the illnesses or injury. This calls for collaboration between the physician and supervisor, during which the full knowledge of the workplace and the compensation recipient's physical state can be known and assessed with the intent of developing a plan for the compensation recipient's future return to work. Physicians need to understand the employee's job function and explore the possibilities of modified duties. Likewise, supervisors should have some sense, from the medical community, whether light or modified duties would be feasible for the employee, and if so they should establish them.

This will enable the employee to return soon and with a positive perspective on returning to work (LRP, 2005).

In addition to the physical state of the compensation recipient, his or her mental state is also important. Many employees feel shame about being disabled, even for a short period. The longer the compensation recipient is away from the workplace, the more they withdraw from their social involvement, which will impede reintegration in the workplace when the time comes. With that in mind the likelihood of returning to work is largely a function of three major factors:

- Time spent on leave
- Interaction among the stakeholders discussed above in the literature review
- Quality of life factors as opposed to type of disorder or disability

It is often said that "the best place to help an injured worker is at work." This is true not only from the emotional sense but the physical sense. At home compensation recipients often spend time alone, sitting or lying, with little or no physical activity. While at work the compensation recipient receives the emotional support of his or her friends, and they also can move around and focus their mind on things other than the way they feel. Most disabilities require some workplace adjustments or "job site modifications." These normally fall into three categories (LRP, 2005):

- Site adjustment – these include changing the layout of the work area.
- Job restructuring – involves changing the employee work hours, adding rest periods to his daily schedule, having him trade jobs with other workers, or limiting or modifying his duties.
- Ergonomic tools - modified hand tools equipment and appliances designed with ergonomics in mind.

These adjustments and modifications must be considered in the agencies' selection of the return to work target group. The second step in the return to work process is to identify the major categories that must be examined to rank the likelihood. The team identified four major categories to rank the likelihood of the compensation recipient returning to work. Those are:

- Employee interest in returning
- Leave already was taken
- Employee's physical condition
- Reassignment factors.

With those identified, the next step is to identify measurable criteria that can be used to score a worker's likelihood of returning to work. The team identified 25 measurable criteria that can be used, see Table 7. These criteria fall into the four general categories that are in column one of Table 7.

Data must be compiled for each compensation recipient on long-term workers' compensation and added to the revised database that will enable scoring of each criterion listed in Table 7. This table is broken into four parts for better viewing.

Area of Concern	No.	Question / Criterion	Score (0 or 1)
Employee Interested in Returning	1	Employee communicated to supervisor interest in returning to work.	
	2	Discussion held between employee and supervisor regarding return to work.	
	3	Physician observes employee expressing an interest in returning to work.	
	4	Since leave was first taken, there has been significant contact by the employee to the supervisor or coworkers on the work-related subject matter.	

Table 7, Part 1 – Scoring of Long-Term Worker's Compensation for Likelihood of Return

Table 7 is filled out using the information found in the compensation recipient's information in the database. The EHS or workers' compensation specialist would go to question one on the table, which is "1."

Area of Concern	No.	Question / Criterion	Score (0 or 1)
Leave Taken (more points for shorter duration)	5	The employee left work within ten years of the scoring of these criteria.	
	6	The employee left within three years of the scoring of these criteria.	
	7	The employee left work within one year of the scoring of three criteria.	
	8	Less than six months have transpired since any major-medical action on the employee's condition (e.g. less than six months after hospital release)	
	9	Less than two months have transpired since any major-medical action.	

Table 7, part 2 – Scoring of Long-Term Worker's Compensation for Likelihood of Return

The employee communicated to supervisor interest in returning. The EHS or worker's compensation specialist then goes to the database and checks that column to find the answer to this question. If the answer is "yes," the compensation recipient did communicate to his or her supervisor an interest in returning to work, one point would be scored (a "1" would be placed in the right-most column of the table for the criterion). If the compensation recipient did not communicate an interest to return to work to his or her supervisor, then no point would be given.

Area of Concern	No.	Question / Criterion	Score (0 or 1)
Physical Conditions	10	The physician recommends employee can return to at least light or part-time duties.	
	11	The doctor says return to work is a possibility (This is a weaker condition than number 10; if number 10 applies, then this condition applies as well).	
	12	The employee is not bed-ridden.	
	13	Continuous monitoring in a facility, or by a nurse at home, is not needed.	
	14	The employee is not connected to medical immobile equipment to sustain him or her.	
	15	The employee is not cognitively impaired to an extent that would preclude performing light duties.	
	16	Reasonable accommodation is possible for part-time work or work from home.	
	17	Employee is experiencing improvements and is awaiting or undergoing treatment to improve his or her condition (e.g. physical therapy)	
	18	The employee is not experiencing chronic pain.	
	19	According to the physician, the condition of the employee is not likely to worsen overtime substantially.	
	20	Additional medical conditions, as evidenced, for example, by extended medical leave before the injury, are not expected to contribute to the disability.	

Table 7, part 3 – Scoring of Long-Term Worker's Compensation for Likelihood of Return

The next step is that the entire table is scored for each compensation recipient in the following manner, with points tallied and the total score evaluated.

- High Probability of Return – for scores of 20-25 points;
- Moderate Probability of Return – for scores of 15-19; and
- Low Probability of Return – for scores of 0-14.

Area of Concern	No.	Question / Criterion	Score (0 or 1)
Re-assignment Factors	21	Work can be assigned similar in nature to work performed before the injury.	
	22	If similar work cannot be assigned, new work could be associated with a similar salary level (so that the new work would not be demeaning)	
	23	The employee has not relocated since the injury to an area where new work cannot be assigned by the Department of Commerce.	
	24	The employee has received favorable performance evaluations before the injury (indicating potential motivation to return).	
	25	A position can be established where the employee can interact with prior coworkers (and would thereby experience less isolation in the new position).	

Table 7, part 4 – Scoring of Long-Term Worker's Compensation for Likelihood of Return

The desirability of a return to work is then determined partially by these scores, in combination with an assessment of the general salary level and the capacity to perform duties. For obvious reasons, employers have a greater interest in returning high paid compensation recipients to work than low paid compensation recipients. Employers are also more interested in returning compensation recipients who can work at full, or near full capacity instead of a compensation recipient whose work hours would be substantially less than full time. The reason is basic economics.

The next step of the process is to use Table 8 to determine the target group that will determine in what order the compensation recipients are returned to work.

Target Group (in descending order of preference)	Likelihood of Return	Relative Salary Level (High or low, above or below mean for organization employees)	Work Capacity (High = Full Time or near Full Time; otherwise low
Target Group 1	High Probability	High Salary	High Capacity
Target Group 2	High Probability	High Salary	Low Capacity
Target Group 2	High Probability	Low Salary	High Capacity
Target Group 3	Moderate Probability	High Salary	High Capacity
Target Group 4	Moderate Probability	High Salary	Low Capacity
Target Group 4	Moderate Probability	Low Salary	High Capacity

Table 8 – Definition of Target Groups for Bringing Employees Back to Work

To explain how Table 8 works, consider the example of "a compensation recipient on long-term workers' compensation leave who is considered to have a high probability of return, a high salary, and is expected to return to work in a high capacity." Now you look for those traits in each of the three columns in Table 8. Each of the three columns lines up to be a Target Group 1 in the far-left column of the table. This means that this compensation recipient is in Target Group 1. Resources are then focused on determining the primary target group for each compensation recipient. This allows for a conscious decision to devote resources on the compensation recipients in Target Group 1 who have the best chance of returning to work in a cost-effective manner. Only after all Target Group 1 compensation recipients have been addressed should the focus be shifted to compensation recipients from Target Group 2, who, although they have a high probability of return, either have a high salary and low capacity or low salary and high capacity and do not provide the return on investment that Target Group 1 compensation recipients do. Once all compensation recipients in Target Groups 1 and 2 have been addressed, resources should be focused on Target Group 3. If resources are available after addressing Target Groups 1 through 3, the focus should be shifted to Target Group 4.

Summary:

The first step in any improvement in workers' compensation is for the employer to demonstrate a commitment to both supervisors and compensation recipients that it expects, and is committed to, progress in this area. However, that will only go so far without a plan to actually make progress. This paper identified the need for a methodology for ranking desirability of a return to work based on measurable criteria. It elaborated on the information gathered in a literature search which led to 25 measurable criteria that can be used to measure the potential for success in a return-to-work effort. Furthermore, the paper broke these criteria into the four key areas. The result is a methodology for ranking desirability of a return to work that is supported by measurable criteria that can be used to measure the potential for success in a return-to-work effort. The method in this paper has never been tried and at this point is only a concept or idea. The author has spent considerable time on this topic since May of 2005 and believes that this method should be tested in a controlled environment to determine its efficacy. When this is done, this method has the potential to reduce long-term workers' compensation claims by as much as 50% in most organizations.

SUMMARY

The goal of this book is to acquaint you with the basics of risk management and the specifics of safety risk management. Risk management started in the financial sector but moved into a variety of other sectors. The method I focused on in this book was safety. This type of risk management can be used to prevent injuries, illnesses, and damage to the environment.

I addressed how to determine the probability that an event might occur compared with the severity of an event. It is important for you to understand both concepts to get a valid identification of the hazard.
Risk management is a process that begins in the planning phase of any activity, continues through the execution, and ends with activity close out. It is an iterative process where feedback is always informing the process of additional risks identified and status of control measures.

I gave you some insight into various types of risk management and then settled on safety risk management. After that, I included a few chapters to show how to adapt the safety risk management process to a variety of situations. In the bibliography, I include a list of sources that I used to write this book. I encourage you to read these books and articles if you want to know more about the subject of risk management. I cannot cover everything in this short book, but I have tried to give you the most valuable information. If you would like additional information about this or any other safety subject, please go to my blog at http://fredefanningauthor.com/.

GLOSSARY

Controls – actions taken to eliminate or control hazards or reduce their severity. There are three primary controls:

  Educational – includes individual and organizational

  Physical – includes barriers, signs, and controls

  Avoidance – prevention

Deliberate Risk Management - application of the safety risk management process.

Hasty Risk Management - a quick mental use of the risk management process during an activity in progress.

Hazard – any real or potential condition that can cause injury, illness, or damage to the environment.

Initial Risk – the level of risk identified before controls have been identified and selected.

Probability – the likelihood that an event will occur.

Residual Risk – the level of risk remaining after controls have been identified, selected, and implemented.

Risk – the probability of exposure to injury, illness, or damage to the environment from a hazard.

Risk Assessment – identification and assessment of hazards (first two steps of the five-step risk management process).

Risk Decision – decision to accept or reject the risk associated with an activity made by individuals responsible for the activity.

Risk Level – the probability or the likelihood an event will occur and the severity or the expected consequence regarding the degree of injury, or damage to the environment.

Risk Management – the process of identifying, assessing, and controlling hazards.

Severity – expected consequences of an event regarding the degree of injuries or illnesses or damage to the environment.

BIBLIOGRAPHY

A Guide to the Project Management Body of Knowledge (PMBOK), Fifth Edition, The Project Management Institute (PMI).

Akel, Philip, Brian Brown, Steven Payson, John Pierson. Strategic Efforts to Maximize the Return to Work of Worker's Compensation Recipients in the US Department of Commerce, July 2006.

Coleman, Thomas S. *A Practical Guide to Risk Management*, Research Foundation of the CFA Institute, 2011, New York, USA.

Crockford, G. Neil. *The Bibliography and History of Risk Management: Some Preliminary Observations*, April 23, 1982. Retrieved on December 15, 2014, from https://www.genevaassociation.org/media/219919/ga1982_gp7(23)_crockford.pdf

Department of Commerce Supervisors Workers' Compensation Handbook, version 1.0. Retrieved from http://ohrm.os.doc.gov/s/groups/public/@doc/@cfoasa/@ohrm/documents/content/prod01_001248.pdf on February 25, 2008.

Dionne, Georges. *Risk Management: History, Definition, and Critique*, March 2013. Retrieved on December 15, 2014 from https://www.cirrelt.ca/DocumentsTravail/CIRRELT-2013-17.pdf

Evans, Doug, and Fanning, Fred. "Risk Management of Wash Rack Operations." "Countermeasures." Volume 20, Number 1, Jan 99:10-11.

Fanning, Fred. "Risk Management for Emergency Operations. 2002 American Society of Safety Engineers Professional Development Conference Proceedings. Jun 02: 71-79.

Fanning, Fred. "A Model Safety Program for Operations Other Than War (OOTW)." Center for Army Lesson's Learned - News from the Front. Jul-Aug 00: 10-15.

Fanning, Fred. "Risk Management for Emergency Operations." 2002 American Society of Safety Engineers Professional Development Conference Proceedings, Jun 02: 71-79.

Fanning, Fred, and Carter T Boggess, Jr. "Risk Management of US Army Rear Detachments. *Perspectives*, Volume 9, Number 3, 2010.

Federal Employees Compensation Act Fact Sheet. Retrieved from http://www.dol.gov/esa/regs/compliance/owcp/fecafact.ht m on February 25, 2008.

Field Manual (US Army) 100-14, Risk Management Program, Washington, USA, 1999.

Injury Compensation for Federal Employees. Department of Labor, September 25, 2007. Retrieved from http://www.dol.gov/esa/regs/compliance/owcp/DFEC%20Folio/ag encyhb.pdf, on February 25, 2008.

"Maximizing Return to Work in the Federal Sector: How to Design, Implement, and Maintain a Successful Program." 2005, LRP Publications, Palm Beach Gardens, FL, USA

"Operational Risk Management: How Best-in-Class Manufacturers Improve Operating Performance with Proactive Risk Reduction." Aberdeen Group, March 2013.

Proper, Ken. "History of Risk Management for Public Sector SH&E Professionals." *Perspective Newsletter* Volume 7, Number 2, 2007.

Proper, Ken. History of Risk Management for Public Sector SH&E Professionals–Part 2–l'Homme Moyen. *Perspective Newsletter*, Volume 7, Number 3, 2007.

Proper, Ken. "History of Risk Management for Public Sector SH&E Professionals: The Avant-Garde Movement." *Perspective Newsletter*, Volume 8, Number 1, 2008. Kaplan, Stanley, and Garrisck, B. John. "On the Quantitative Definition of Risk." Retrieved on March 7, 2015, from http://onlinelibrary.wiley.com/doi/10.1111/j.1539-6924.1981.tb01350.x/abstract.

"Title 5-Government Organization and Employees, Part III--Employees, Subpart G-Insurance and Annuities," Chapter 81- Compensation for Work Injuries. Retrieved from http://www.access.gpo.gov/uscode/title5/partiii_subpartg_chapter81 _.html on February 25, 2008.

US Army Field Manual 100-14, Risk Management, 1999.

US Army Field Manual 101-5, Staff Organization and Operations, 2001.

ABOUT THE AUTHOR

Fred Fanning spent over 20 years as a safety professional with the US Government culminating as the Director of the Office of Occupational Safety and Health for the US Department of Commerce. Fred is also an independent author. His published work includes two editions of the peer-reviewed book *Basic Safety Administration-A Handbook for the New Safety Specialist*. Fred also authored two editions of the peer-reviewed chapter *Safety Training and Documentation Principles* published in the bestselling *Safety Professional Handbook* and the *Safety Professional Handbook Management Applications*. He coauthored the chapter *Safety Training* with Christine Fiori, Ph.D., PE, published in the bestselling Construction Safety Management and Engineering, second edition edited by Darryl C. Hill, Ph.D., CSP. He also authored several other books on occupational safety and health. Fred has over 40 technical articles published. Fred has written for several publications including the Project Management Institute's Government Community of Practice Magazine; the American Society of Safety Engineers' Professional Safety Journal; and the American Society of Safety Engineers' Perspectives Newsletter.

Fred holds the Certified Facility Manager (CFM) certification from the International Facility Management Association, Project Management Professional (PMP) certification from the Project Management Institute and the Leadership in Energy and Environmental Design (LEED) Green Associate from the Green Building Certification Institute. He held the Certified Safety Professional certification from 1995 through 2010. Fred earned masters' degrees from National-Louis University and Webster University.